what about...
planet
earth?

what about...
planet earth?

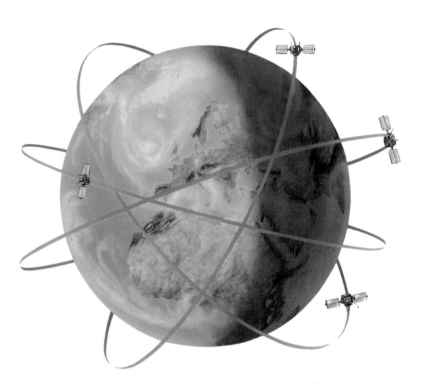

Brian Williams

MC
PUBLISHERS

This 2009 edition published and distributed by:
Mason Crest Publishers Inc.
370 Reed Road, Broomall, Pennsylvania 19008
(866) MCP-BOOK (toll free)
www.masoncrest.com

Library of Congress Cataloging-in-Publication data is available

What About...Planet Earth
ISBN 978-1-4222-1564-7

What About ... - 10 Title Series
ISBN 978-1-4222-1557-9

Printed in the United States of America

First published in 2004 by Miles Kelly Publishing Ltd
Bardfield Centre, Great Bardfield Essex CM7 4SL

Copyright © 2004 Miles Kelly Publishing Ltd

Editorial Director Belinda Gallagher
Art Director Jo Brewer
Senior Editor Jenni Rainford
Assistant Editors Lucy Dowling, Teri Mort
Copy Editor Rosalind Beckman
Design Concept John Christopher
Volume Designers Jo Brewer, Michelle Cannatella
Picture Researcher Liberty Newton
Indexer Helen Snaith
Reprints Controller Bethan Ellish
Production Manager Elizabeth Brunwin
Reprographics Anthony Cambray, Stephan Davis, Liberty Newton, Ian Paulyn

All images from the Miles Kelly Archives

CONTENTS

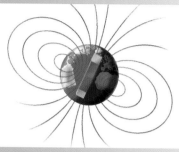

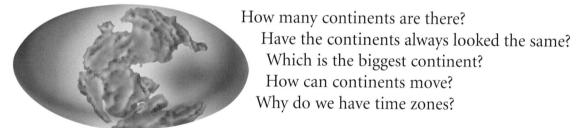

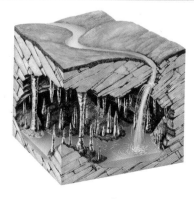

6 CONTENTS

Volcanoes and Earthquakes

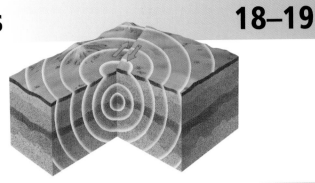

What makes a volcano erupt?
What made the loudest bang ever?
Why are some volcanoes not dangerous?
What sets off an earthquake?
What is a tsunami?

Landscapes

What is the highest mountain on land?
How do rivers shape the land?
How can weather change a landscape?
What is the world's biggest canyon?
What starts an avalanche?

Rivers and Lakes

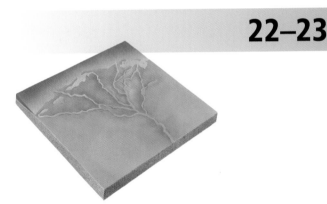

Which is the deepest lake?
How did the Great Lakes get their name?
Which river has the largest volume of water?
Where would you find a delta?
Can a river flow backwards?
What makes a waterfall?

Oceans

What causes tides?
How many oceans are there?
Which is the biggest ocean?
What is it like on the ocean floor?
Why is the Dead Sea so salty?
Could you drink a melted iceberg?

Deserts

Which is the biggest desert?
Are all deserts sandy?
Can sand dunes move?
What is an oasis?
How do desert animals and plants survive?

Forests

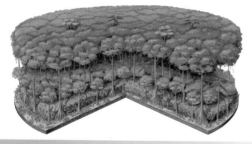

Are there different types of forest?
Which is the world's biggest rain forest?
What starts a forest fire?
Why do some forest trees shed their leaves?
What is meant by a sustainable forest?

Atmosphere

How many layers are there in the atmosphere?
How does the atmosphere protect us?
Why does the sky look blue?
What causes the northern and southern lights?
Where is the air coldest?

Weather and Climate

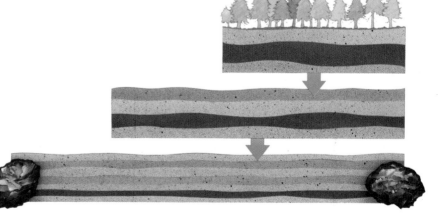

What are the highest clouds in the sky?
How deep can snow be?
Where is the thickest ice?
What is a hurricane?
What causes lightning?
When might you see a rainbow?

Earth Resources

What are raw materials?
How is water used to make electricity?
Why is a diamond like a lump of coal?
What is coal?
Where is the world's richest goldfield?
What is the 'greenhouse' effect?

There is no other planet like our Earth—so far as we know. The Earth is unique in the solar system because it is the only planet with life. The Earth has a vast array of environments, from scorching deserts to misty rain forests, in which living things have thrived for millions of years. The Earth is still changing, reshaped by natural forces and, increasingly, by human activity.

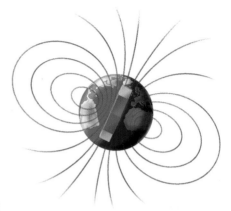

⬆ *The Earth's magnetic field stretches out invisibly into space.*

⬆ *The distance around the Earth at the poles is slightly less than the distance around the Earth's middle, the equator. Although the planet looks round, it has a bulge just south of the equator.*

When was the Earth formed?

The Sun and its nine planets formed at around the same time, about 4,600 million years ago. The Earth grew from a whirling cloud of gas and dust in space, which became squashed together by gravity. Most of the material in the cloud came together in the middle to make the Sun. The debris left over formed gas balls and rocky lumps: the planets. One of those lumps was the Earth.

➡ *The young Earth was a violent planet, shaped by storms and volcanoes.*

How is the Earth like a magnet?

The Earth's magnetism is made by the Earth's electrical currents, generated by movements inside the planet. It has a magnetic field stretching far out into space. Like all magnets, the Earth has north and south poles where the magnetism is strongest.

How big is the Earth?

The Earth is the fifth biggest of the Sun's planets, with a circumference of 24,902 miles and weighing about 6,000 million million million tons. It is a ball of rock, mostly covered by water, wrapped in a thin, protective layer of gases—the air. Yet the Earth is tiny compared with the planet Jupiter, which is more than 300 times bigger.

Earth's vital **statistics**

Discovering Earth
The Earth was first measured in about 200 BC by a Greek mathematician named Erathosthenes. Using geometry, he measured the angle of the Sun's rays (shown by a shadow) at two places a known distance apart. Eratosthenes worked out that the Earth must be around 28,500 mi. in circumference—pretty close, compared to the modern figure of 24,902 mi. Nowadays, the Earth can be measured by satellites using lasers.

In a year the Earth travels once around the Sun—a distance of over 584,000,000 km. This means that when astronomers look at the star through a telescope, they are actually looking back into the past—seeing the star as it was four years ago.

⬇ *The Earth spins as it travels in its year-long orbit around the Sun. Because it spins faster at the equator than at the poles, the Earth is not a perfect sphere. One rotation takes 24 hours, or one day.*

⬇ *Seen from space, the Earth is a planet of oceans. Only about 29 percent is land.*

Why is the Earth a watery world?

Earth is a watery world because about 71 percent of it is water. The water on the planet is in the oceans, as ice at the poles and on mountain tops, in lakes and rivers, and in water vapor in the atmosphere that falls as rain. The rainiest places on Earth are near the equator on the coast or on islands. In parts of West Africa and the Amazon region of Brazil it rains almost every day.

Pacific Ocean

➡ *Oceans contain 97 percent of the Earth's water and the three biggest (Pacific, Atlantic, and Indian) together cover an area of 135 million sq. miles.*

Atlantic Ocean

Indian Ocean

How is the Earth moving?

The Earth moves in three ways: on its own axis, through space, and as part of the solar system. First it spins on its own axis—an imaginary line from pole to pole. Second it races through space as it orbits the Sun, held in position by the Sun's gravity. Third it is part of the solar system, and moves through space as the Milky Way galaxy, which contains the Earth, rotates at about 150 miles/sec.

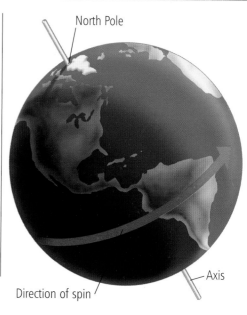

North Pole

Axis

Direction of spin

⬆ *The Earth spins on its axis—an imaginary line between the poles.*

⬅ *The Sun's rays provide the Earth with light and warmth. The planet is tilted at an angle, and as it spins in orbit around the Sun, different regions get varying amounts of sunlight. The part tilted away from the Sun has a cool winter season; the part tilted towards the Sun (here the southern hemisphere) is warmed up more and so has summer.*

Earth's **vital statistics**

Equatorial circumference	24,902 miles
Polar circumference	24,860 miles
Surface area	196.8 million sq. miles
Land area	29 percent
Water area	71 percent
Highest temperature	136.4°F in Libya, North Africa
Lowest temperature	−129.3°F in Antarctica
Most abundant chemicals	oxygen 47 percent, silicon 28 percent, aluminum 8 percent, iron 5 percent

The Earth is made of rocks. Geologists (scientists who study the Earth's rock history) can uncover the prehistory of the Earth by studying the rocks, which are laid down in layers rather like a gigantic sandwich. Rocks provide us with useful materials, such as coal for fuel and limestone to make cement. They also contain clues to life long ago in the form of fossils—the hardened remains of long-dead animals and plants.

What is inside the Earth?

The Earth's crust rests on a layer of hot, partly molten rock called the mantle, which in turn surrounds the two cores. The core or center of the Earth is a solid ball of very hot rock, about 4,000 miles beneath the surface and under enormous pressure. The core is too hot and solid for us to drill down to. The deeper down, the hotter it gets—more than 7,200°F at the Earth's core.

What is the Earth's crust?

The Earth's rocky skin is the crust, which is thickest (up to 25 miles) beneath "young" mountain ranges. The crust beneath the oceans is much thinner, between 3 and 7 miles. The rocks of the continents are much older than the rocks under the oceans.

↑ *The Giant's Causeway in Ireland is a formation of basalt, the most common volcanic rock in the Earth's crust.*

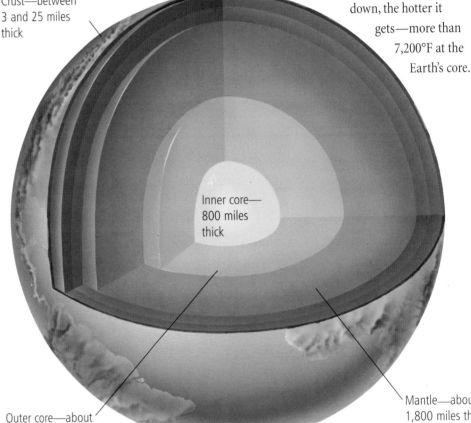

Crust—between 3 and 25 miles thick

Inner core— 800 miles thick

Outer core—about 1,400 miles thick

Mantle—about 1,800 miles thick

How **rocks are formed**

Different rocks

The three main kinds of rocks are igneous, sedimentary, and metamorphic rocks. Igneous rocks are made from molten rock called magma, deep inside the Earth. Magma is so hot, over 1,800°F, that it is molten. It is also crushed by enormous pressure. When magma gets pushed to the surface by volcanic action it cools to form igneous rocks.

Sedimentary rocks such as shale and sandstone are made by the action of wind and water, which grind other rocks into sand and mud, carried by rivers until they are deposited as sediments. Sediment piles up in layers, and is squeezed hard by the pressure of layers on top until it becomes rock.

Metamorphosis means "change" and metamorphic rocks are changed by chemical action, heat, or pressure into another form of rock. This can happen when magma pushes through them or when the Earth's crust moves beneath mountain ranges. Limestone, for instance, becomes marble when subjected to these kinds of changes: A sedimentary rock becomes a metamorphic rock.

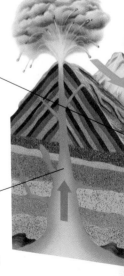

Igneous rock is formed when lava cools

As volcanoes erupt, hot magma is released from inside the Earth

What are fossils?

Fossils are the hardened remains of dead animals and plants. They are found in rocks that were once soft sand or mud, such as sandstone, often when ancient rocks are exposed by wind or rain weathering, or by quarrying and mining. Shells, bones, and teeth are most likely to end up as fossils rather than soft parts, which decay. Sometimes scientists find a whole skeleton, which can be removed bone by bone and carefully reconstructed.

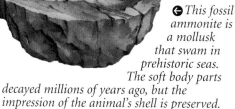

This fossil ammonite is a mollusk that swam in prehistoric seas. The soft body parts decayed millions of years ago, but the impression of the animal's shell is preserved.

Fossil of a Tyrannosaurus rex *skull, a dinosaur that lived on Earth about 65–70 million years ago.*

This ancient trilobite, a shellfish, would have died millions of years ago. Over time only its bones and shell remain, buried by minerals in the seawater to become a fossil.

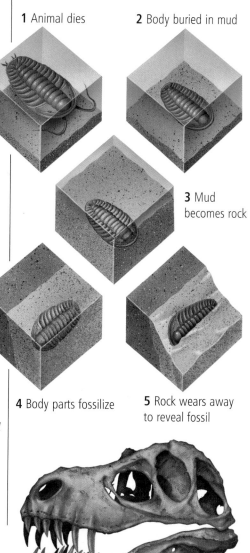

1 Animal dies

2 Body buried in mud

3 Mud becomes rock

4 Body parts fossilize

5 Rock wears away to reveal fossil

What are rocks made of?

Rocks are made of minerals. Most rocks are what geologists call aggregates—that is, combinations of several minerals. There are three kinds of rocks, made in different ways. The most common kind are sedimentary rocks.

Chalk is a form of limestone made from the shells of tiny single-celled sea animals called foraminiferans. Limestone is a type of sedimentary rock.

Which is the most common element?

The most common element in the universe is hydrogen, but on Earth it's oxygen, which accounts for about 47 percent of the planet's mass. Elements are substances that are made of only one kind of atom. All matter in the universe is made of elements.

Rain and wind breaks down rocks

Rock fragments, broken by the weather, are washed down into the sea

Sedimentary rock forms on the seabed from rock debris

Igneous rocks are made deep inside the Earth from cooled magma. Metamorphic rocks are made from other rocks, changed by heat and pressure, as in a volcano. Sedimentary rocks are formed from the worn and squashed fragments of other rocks and living things.

Rock materials are continually recycled to make new rocks.

Scientists study the Earth's history to understand how it has changed, and is still changing, as natural forces reshape the landscape.

Continents are landmasses with water all around them, or almost all around them. The continents contain landscape—such as mountains, rivers, lakes, deserts, grassy plains, forests, and cities. The continents are made of very old rocks, dating back some 3,800 million years. Yet, although they are so massive, they are not fixed in place but are drifting very slowly.

➲ *This is what Pangaea, the original supercontinent, may have looked like. Scientists believe the present continents have reached their shape and position by a process of fracture and drift over millions of years.*

PANTHALASSA OCEAN

Pangaea

TETHYS SEA

Have the continents always looked the same?

No, all the continents once formed one huge landmass, which scientists call Pangaea. This was 280 million years ago. Over time, the supercontinent broke up into two smaller but still huge continents—Laurasia and Gondwanaland. Laurasia included North America, Europe, and part of Asia and Gondwanaland contained South America, Africa, Australia, Antarctica, and India. Later, these fractured and the pieces drifted apart to form the continents as we know them today.

How many continents are there?

There are seven continents: Africa, Antarctica, Asia, Europe, Oceania, North America, and South America. Each of the continents encompasses various countries and bodies of water. From the map (right), you can see how the outlines of South America and Africa look as if they might fit together—suggesting that they were once joined together.

➲ *The seven continents include Antarctica, which has land beneath its thick ice, but not the Arctic, which is mostly frozen ocean.*

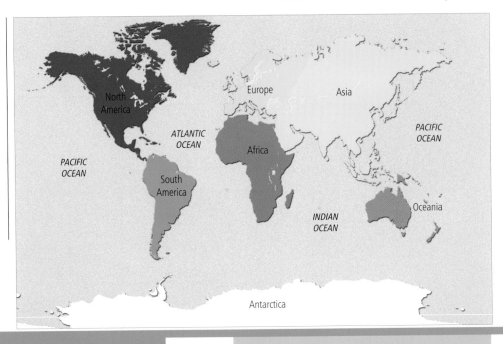

North America

Europe

Asia

ATLANTIC OCEAN

Africa

PACIFIC OCEAN

PACIFIC OCEAN

South America

INDIAN OCEAN

Oceania

Antarctica

Continental **facts**

Continental drift

Continental drift can be explained by the theory of seafloor spreading. Hot molten rock is pushed up beneath the seafloor, where the crust is thinnest, and forced into cracks in the midocean ridges. As the molten rock cools, it hardens, pushing the ocean floor and the continents away from the ridges.

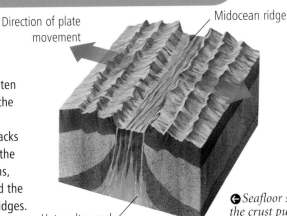

Direction of plate movement

Midocean ridge

Hot molten rock

Amazing **facts**

• By 135 million years ago, South America was drifting away from Africa.

• By 100 million years ago, India, Australia, and Antarctica were drifting away from Africa, too.

• And at the same time, North America was moving away from Europe.

• The plates are typically about 60 miles thick.

➲ *Seafloor spreading helps to move the continents. Currents beneath the crust push molten rocks up to the midocean ridges and into the ocean valleys. As the melted rock cools, it makes a new seafloor.*

Which is the biggest continent?

Asia is by far the biggest continent. The total land area of Asia is 17 million sq. miles, which is four times bigger than Europe and nearly twice as big as North America. Asia includes the biggest country by land area (Russia) and the two biggest by population (China and India).

How can continents move?

The Earth's crust is made of curved rocky plates, which float like pieces of gigantic jigsaw on the molten layer of hot rocks in the mantle. There are seven large plates and some 20 smaller ones, and they move very slowly (between 0.4 and 4 inches a year) on currents circulating within the mantle. Over millions of years, the continents that rest on top of these plates move, too.

🔼 *Ocean plates can be pushed down beneath other plates, producing unstable movements beneath continents. Rocks are folded by pressure and magma is forced up through volcanoes.*

Why do we have time zones?

Time is not the same all around the Earth because the Earth turns once on its axis in just under 24 hours. When one spot on the Earth is in sunlight and so has day, another place on the far side is in shadow and has night. The world has 24 different time zones, and the clock time in each zone differs by one hour from that in the next. The United States and Canada are such big countries that they encompass six time zones.

🔽 *When the time is 12 noon in London, U.K., it is 7 a.m. in New York, U.S.A. (five hours earlier) and 11 p.m. in Wellington, New Zealand (11 hours later).*

Asia

PACIFIC OCEAN

INDIAN OCEAN

🔙 *This view of the globe shows Asia, which extends into Europe and is the biggest of the continents.*

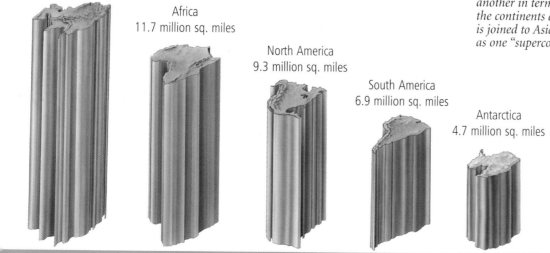

7 a.m. in the U.S.A.

Noon in England

11 p.m. in New Zealand

Asia
17 million sq. miles

Africa
11.7 million sq. miles

North America
9.3 million sq. miles

South America
6.9 million sq. miles

Antarctica
4.7 million sq. miles

Europe
4 million sq. miles

Oceania
3.3 million sq. mies

🔙 *The continents are shown here in relation to one another in terms of land area. Asia is the largest of the continents and Oceania is the smallest. Europe is joined to Asia, so the two are sometimes treated as one "supercontinent" called Eurasia.*

14 MAPS AND GLOBES

Geographers (scientists who study the Earth and its features) rely on maps. The word "geography" comes from a Greek word meaning "description of the Earth." A map is a small picture of a large area, and is drawn to scale (for example 1 inch on the map might represent 1,000 miles of land on the ground). Most maps are flat, but a globe is spherical, like the Earth. Modern maps are very accurate, made with the aid of computers and photographs that have been taken from aircraft and satellites in space.

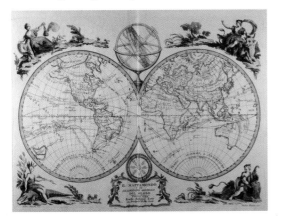

⊙ *The first reasonably accurate maps of the world were made in Europe in the 1500s. Lines of latitude and longitude are marked in degrees. On a map, 1° is one 360th of a circle.*

Who made the first maps?

About 5,000 years ago people in Egypt and Babylonia made drawings to show who owned which bit of land and where rivers were. The oldest map in existence is a clay tablet found in Iraq that has what may be a river valley scratched on it. The first maps to show lines of latitude and longitude, to fix a position, were made by the ancient Greeks about 2,000 years ago.

Where are the tropics?

The tropics are the regions of the Earth that lie immediately north and south of the equator, an imaginary line around the middle of the Earth. The northern region is the tropic of Cancer, the southern region is the tropic of Capricorn. Each tropic is about 1,600 miles wide and here the Sun shines almost directly overhead at noon. The tropics are each approximately 1,600 miles from the equator. On a map, the tropics lie at a latitude of 23° north and south of the equator.

⊙ *The equator is an imaginary line around the middle of the Earth, with the tropics north and south of it.*

Tropic of Cancer

Equator

Tropic of Capricorn

Mapmaking

Projections

A flat map of a sphere like the Earth cannot be accurate unless certain adjustments are made. If you peel an orange carefully, you will discover that the peel won't lie flat on a table without breaking. Maps are drawn in a way that makes one feature (such as land area) accurate, but another feature (shape, for instance) less so. These different ways of mapmaking are known as projections. There are various kinds of map projection. One is named Mercator after a Flemish mapmaker named Gerardus Mercator (1512–94). This projection shows the correct direction between two points, because the lines of latitude and longitude are correct. It makes landmasses look wrong though—Greenland looks the same size as North America, which is actually really much larger.

➡ *A conic projection of a globe projects the lines of latitude and longitude into a cone shape, which can then be flattened to give a picture of the wide landmasses, such as the U.S.A. and Russia, as accurately as possible.*

What are latitude and longitude?

A network of lines across a map, forming a grid. Lines of longitude are drawn from north to south, while lines of latitude are drawn from east to west. The lines make it easier to locate any spot on the map. The equator is the line of 0° latitude. The line of 0° longitude runs through Greenwich in London, England and is known as the prime meridian.

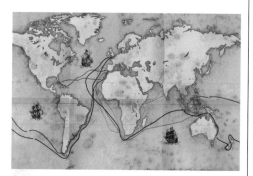

⊘ *Early navigators used a sextant to measure the height of the Sun above the horizon to help them work out latitude. Finding longitude become possible in the 1750s with the invention of accurate clocks for use at sea.*

⊙ *Cook's voyage (shown in blue) and Magellan's voyage (shown in green) sailed around the Oceanic islands for the first time.*

Why do people never need to get lost today?

Global positioning system (GPS) satellites in orbit around the Earth can inform travelers where they are to within a few feet. The satellites send out radio signals that are picked up by a computer on an airplane, ship, or car; three or more "fixes" give a precise position. Satellite navigation began in 1960 with the U.S. Transit satellite. The more advanced Global positioning system became operational in 1995.

Why is Australia not shown on early maps?

Because until the 1600s no one in the Northern Hemisphere knew it was there. Chinese and Indonesian sailors may have been the first outsiders to visit Australia, after the aboriginal people settled there some 40,000 years ago. European sailors discovered Australia by accident when straying off course on voyages to Asia. The discovery of the islands that make up the continent of Oceania was only made from the 1500s onward when sailors such as Ferdinand Magellan and Captain James Cook sailed the southern hemisphere. However, the islands were already inhabited by this time by the Polynesians, Melanesians, and Micronesians, who had settled there after arriving from Asia more than 1,000 years before.

⊙ *Navigation satellites now encircle the Earth, providing data for travellers to be able to find where they are at any given moment.*

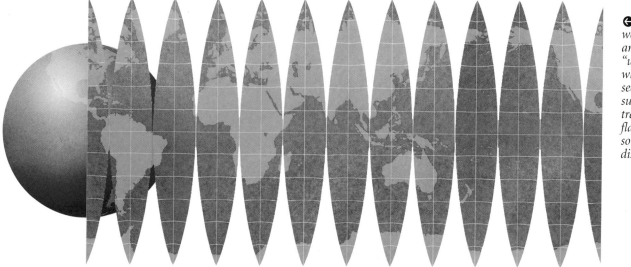

⊙ *If the Earth were an orange and could be "unpeeled," this is what you would see. The curved surface cannot be transformed into a flat map unless some features are distorted.*

The surface of the Earth looks solid, but underground in some places there are holes or caves. Most were hollowed out of soft rock by water trickling down from the surface. Some are as big as football fields! Others contain unusual mineral formations. People who explore caves are known as cavers, spelunkers, or (in the U.K.) potholers.

Where are the world's longest caves?

The world's longest caves are the Mammoth Caves of Kentucky in the United States, first explored in 1799. This cave system has 350 miles of caves and passages, with underground lakes and rivers.

⬆ *Some cave systems contain huge caverns, large enough for people to stand in. Others are cramped and narrow, and can only be crawled through by cave explorers.*

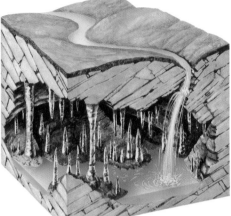

⬆ *Stalactites and stalagmites are made by the slow build-up of minerals in dripping water. They turn a cavern into a subterranean wonderland, full of interesting shapes.*

How can you tell a stalactite from a stalagmite?

Stalactites are mineral formations that look like giant icicles as they hang down from the roof of a cave. A stalactite more than 40 ft. long was measured in a cave in Brazil. Stalagmites grow up from the floor of a cave, as water drips down from the cave roof. A stalagmite more than 100 ft. high was measured inside a cave in Slovakia.

Which caves have the oldest paintings?

Prehistoric people lived in caves for protection against the weather and wild animals. Some caves contain pictures of animals made by these cave dwellers, and the best known are those at Lascaux in France and Altamira in Spain. Similar paintings were found in 2003 at Creswell Crags in Nottinghamshire, England. A few large caves were home to many generations of prehistoric people.

⬇ *Stone Age people drew the cave paintings, such as this bison, at Lascaux in France more than 13,000 years ago.*

Going **under**

Underground
Caves are holes in the ground, usually hollowed out by water wearing away soft rock. But many "potholes" are so narrow that explorers have to crawl on hands and knees, or even swim through flooded sections of cave, using flashlights to penetrate the gloom.

➡ *An explorer inside a cave that has been made from lava from a volcano.*

The biggest cave chamber is called the Sarawak Chamber, in a cave system in Sarawak, Malaysia. It is 2,300 ft. long, has an average width of 980 ft. and is about 230 ft. above the cave floor.

Deep caves **of the world**

Krubera, Georgia, Asia	5,610 ft.
Reseau Jean Bernard, France	5,256 ft.
Shakta Pantujhina, Georgia	4,947 ft.
Sistema Huautla, Mexico	4,839 ft.
Sistema del Trava, Spain	4,727 ft.
Vercors, southeast France	4,170 ft.
Gunung Mulu, Borneo	1,542 ft.
Carlsbad Cavern, USA	1,037 ft.

Stream flows
underground

Underground waterfall

Stalactites

Stalagmites

Limestone
rock worn
away to
form cavern

Water emerges
into a lake

⬆ *Limestone caverns and cave systems are eroded (worn away) by chemical weathering.*

What caves are made by chemistry?

The soft rock in limestone caves is worn away by "chemical weathering." Calcium carbonate in the limestone reacts with rainwater to form a weak acid, which gradually dissolves the rock. Water seeping down the rock forms cracks and potholes, which open up into caverns. The chemical "drips" can create growths, and formations of stalactites and stalagmites. Underground waterfalls, rivers, and lakes form, and water often flows into an open-air lake or river.

Do any animals live in caves?

Caves provide shelter for a number of animals, including bats and birds, such as cave swiftlets of Asia and Caribbean oilbirds. These animals roost in the cave by day (bats) or at night (birds) and come out to hunt for food. Cave swiftlets are hunted by the racer snakes who also inhabit the caves. Many insects also live in caves, and underground lakes are home to various species of fish. Many cave species are blind and so rely on smell, touch, or echolocation (using echoes from sound to judge distances and obstacles) to find their way around in the darkness.

⬆ *Birds such as swifts build their nests in caves.*

⬇ *Rocks inside caves may contain impressions of fossil plants that grew millions of years ago.*

1 Streams enter from the surface and water trickles down through cracks in the stone

2 As the water dissolves the limestone, it hollows out tunnels and caverns

⬆ ➡ *These diagrams show how limestone caves are formed by the action of water.*

3 In time, part of the cavern roof may collapse, and the stream becomes a subterranean river

In parts of the world violent upheavals shake the ground and send fire and smoke into the sky, causing terrible damage and loss of life. These are volcanoes and earthquakes, and they often occur in the same regions. Volcanoes belch smoke, fire, ash, and molten rock. Earthquakes shake the ground with an energy many times greater than that of an atomic bomb.

What makes a volcano erupt?

A volcano erupts when molten rock (called magma) is pushed up from deep inside the Earth and is forced out through the mouth of the volcano. When the magma reaches the air it becomes lava, flowing down the sides of the volcano. Some volcanoes explode violently, hurling rocks, lava, and ash into the air.

What made the loudest ever bang?

The biggest volcanic explosion was in 1883, when Krakatoa Island in Indonesia blew up. The noise was heard four hours later almost 3,000 miles away! The tidal wave Krakatoa produced killed 36,000 people. However, after the Tambora volcano (which struck Indonesia in 1815) 90,000 people were killed by the volcano, the tidal wave, or the famine that occurred afterwards.

⊙ Erupting volcanoes throw smoke, ash, and rocks into the sky.

Why are some volcanoes not dangerous?

Not all volcanoes erupt. Extinct volcanoes can no longer erupt. Dormant volcanoes do so only occasionally. Some just "grumble," trickling out lava and steam. However, active volcanoes erupt often. The most famous is Mount Vesuvius, above the Italian city of Naples. It destroyed the Roman city of Pompeii in AD 79, and will almost certainly erupt in the 21st century.

⊙ *Mount Fuji is Japan's biggest mountain— the result of many eruptions of ash and lava. It is now extinct, and last erupted in 1707.*

Earth in **upheaval**

Famous volcanoes
Some volcanoes are particularly famous, either due to their size, or as a result of having erupted with devastating consequences.

The extinct volcano Aconcagua in Argentina is the highest volcano in the western hemisphere.

Mount Etna in Sicily, Italy, erupted in 1669, killing 20,000 people.

Mount Vesuvius in Naples, Italy, destroyed Pompeii and Herculaneum when it erupted in AD 79.

⊙ *When a volcano erupts, molten magma explodes from the main vent. Ash and lava pour out and flow down the side of the volcano. Gas, dust, and rock "bombs" are thrown into the sky.*

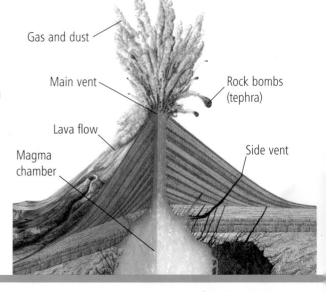

Gas and dust

Main vent

Rock bombs (tephra)

Lava flow

Magma chamber

Side vent

⬆ *During an earthquake, the rocks shift along a fault line in the crust. Here, part of a road has fallen away during a quake.*

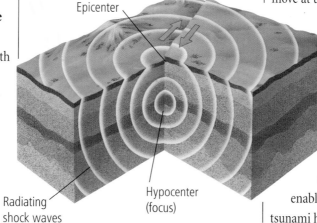

➡ *Tsunamis are large and potentially devastating waves that are caused by undersea earthquakes.*

What sets off an earthquake?

Like volcanoes, earthquakes occur where the rocks of the Earth's crust are put under tremendous pressure by movements of the plates on which the continents rest. Seismologists (scientists who study earthquakes) measure the strength of the shock waves with a seismometer. The earthquake is then graded using the Richter scale, which starts at 1 (slight tremor) and rises above 9 (devastating quake). The world's worst earthquake of modern times was in 1976, when a quake in China measuring 8.2 on the Richter scale led to the deaths of at least 250,000 people.

⬇ *During an earthquake, pressure waves ripple out from the epicenter. Earthquakes are caused as two blocks of rock crust move along a fault line (shown by the arrows) in opposite directions.*

Epicenter

Radiating shock waves

Hypocenter (focus)

What is a tsunami?

Undersea earthquakes can produce huge waves, called tsunamis. These waves move at up to 500 miles/h. but may not be noticed in the open sea. In shallow waters however, the wave builds into a colossal wall of water up to 100 ft. high, which rushes inland drowning everything in its path. A warning system was developed in 1948 in the Pacific Basin, where most tsunamis occur. Several hours' warning can enable areas to be evacuated before the tsunami hits to minimize casualties.

Amazing **volcano facts**

- The biggest volcano is Mauna Loa in Hawaii, with a crater 6 miles wide and 590 ft. deep. More than 80 percent of this volcano is beneath the ocean.

- The highest active volcano is Ojos del Salado in South America at 22,572 ft.

- The most restless volcano is Kilauea in Hawaii, which has been in constant eruption since 1983. Its lava flows have covered more than 39 sq. miles.

- There are more than 800 active volcanoes in the world. The country with the most is Indonesia, which has about 200.

- Tambora in Indonesia erupted in 1815, killing approximately 90,000 people.

Earthquakes of the **20th century**

Place	Date	Richter scale
San Francisco, U.S.A.	1906	8.3
Gansu, China	1920	8.6
Kanto Plain, Japan	1923	8.3
Assam, India	1950	8.6
Aleutian Islands	1957	9.1
Tangshan, China	1976	8.2

Mountains are made by movements of the Earth's rocky crust. The highest mountains are the youngest and are still growing, pushed up by enormous pressure deep inside the Earth. For example, the Himalayas have been built up over the last 40 million years, whereas the Adirondacks in New York, which are over one billion years old, have been worn flat, or reduced to mere hills.

⬆ Rivers meander through the landscape to flow into open water, such as lakes, seas or oceans, shaping the land as they run.

What is the highest mountain on land?

Mount Everest, which rises to 29,035 ft. in the Himalayan range of Asia. The Himalayan range has the world's 20 highest mountains, all more than 26,000 ft. high. An even higher peak, Mauna Kea in Hawaii, rises out of the Pacific Ocean. It rises 33,480 ft. from the ocean floor to its summit, but only 13,796 ft. of this is above water.

How do rivers shape the land?

Over thousands of years a river carves out a course through the rocks and soil, as it flows towards the sea or into lakes. These river courses create valleys and canyons, and where they wind slowly through flat country, they move in snaky curves called meanders. They wear away the bank on one side and pile up silt on the other.

⬇ The highest mountains in each continent.

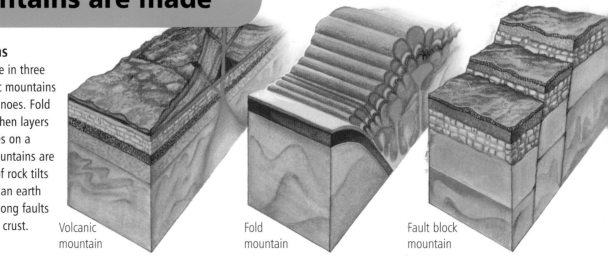

Puncak Jaya
Oceania
16,503 ft.

Elbrus
Europe
18,510 ft.

Kilimanjaro
Africa
19,340 ft.

McKinley
North America
20,320 ft.

Aconcagua
South America
22,835 ft.

Everest
Asia
29,035 ft.

How **mountains are made**

Forming mountains
Mountains can be made in three different ways. Volcanic mountains are pushed up by volcanoes. Fold mountains are made when layers of rock fold like wrinkles on a blanket. Fault block mountains are made when a section of rock tilts or is pushed up during an earth tremor. This happens along faults or breaks in the Earth's crust.

Volcanic
mountain

Fold
mountain

Fault block
mountain

↑ *The biggest canyon on Earth is the Grand Canyon in the U.S.A.—in places it is 1 mile deep.*

What is the world's biggest canyon?

The Grand Canyon is a huge gorge in the Earth's surface, which has been cut by the Colorado River in Arizona, U.S.A. The process of cutting the canyon, 277 miles long and in places more than 18 miles across, has taken millions of years. In some parts of the canyon, rocks of up to two billion years old have been uncovered.

What starts an avalanche?

Avalanches can be started by strong winds, melting snow, loud noises, or even by people skiing on loose snow. An avalanche is an enormous mass of snow that slips down a mountainside. Millions of tonnes of snow fall at speeds of up to 250 miles/h. The biggest avalanches occur in the world's highest mountain range—the Himalayas.

↑ *The fastest avalanches are masses of dry snow. Wet snow moves at a more sluggish pace.*

How can weather change a landscape?

The Earth's landscape is changing all the time because of erosion, which is the wearing away of rocks and soil by "scouring" forces such as wind, water, ice, and frost. In winter, water trapped in cracks in rock freezes, expands, and causes chunks of rock to split off. Heavy rain can quickly wash soil down a slope, especially if there are no trees on the slope to "bind" the soil with their roots.

➔ *The landscape is shaped by the weather often into formations like this rock arch.*

Highest **mountains**

The five highest peaks in the world are all in the Himalayan range in Asia:

Everest	29,035 ft.
K2	28,250 ft.
Kanchenjunga	28,208 ft.
Lhotse	27,890 ft.
Makalu	27,824 ft.

Other **famous peaks**

Cotopaxi (Mexico)	19,347 ft.
Ararat (Turkey)	16,946 ft.
Mont Blanc (France/ Italy/ Switzerland)	15,771 ft.
Matterhorn (Switzerland/Italy)	14,692 ft.
Jungfrau (Switzerland)	13,642 ft.
Fuji (Japan)	12,388 ft.
Olympus (Greece)	9,570 ft.

➔ *Mount Everest was first climbed in 1953 by Sherpa Tenzing Norgay and New Zealander Edmund Hillary, supported by a British Commonwealth team of climbers.*

Water from the oceans is drawn up into the air as water vapor by the warmth of the Sun. It is blown over land by winds, cools to become water droplets, and falls as rain. Rainwater fills up lakes and rivers, and eventually finds its way back to the oceans. All the water on Earth is recycled, over and over again.

4 Rivers flow into ocean

3 Water vapor falls as rain and snow

2 Moisture from ground evaporates

1 Ocean water evaporates

In the water cycle, water is drawn up from the oceans by evaporation and later falls as rain over the land. This is known as precipitation. The rainwater then flows back to the ocean in rivers.

Which is the deepest lake?

Baikal in Siberia, Russia, is 5,315 ft. deep, roughly four times deeper than Lake Superior in North America. Baikal is a very ancient lake, about 25 million years old, and is unique in being home to the world's only freshwater seal. Lake Tanganyika in Africa is the next deepest lake but is only two million years old.

How did the Great Lakes get their name?

Canada and the United States share the five Great Lakes; they got their name because they are the biggest group of freshwater lakes in the world. They are: Lake Superior (the world's biggest freshwater lake), Lake Huron (the fifth largest lake in the world), Lake Michigan (the sixth largest lake), Lake Erie, and Lake Ontario. Canada has the most fresh water (by area)—twice as much as any other country.

Which river has the largest volume of water?

The Amazon River carries more water than any other. It pours 20 billion gal. of water into the Atlantic Ocean every hour. The Amazon also has the most tributaries (rivers that flow into it). Second to the Amazon in volume of water flow is China's Huang He, which is the muddiest river. Another name for Huang He is the Yellow River, so-called because of the color of the silt washed into it.

The Amazon River winds in and out of lush rainforest that grows on either side.

Lake Superior
Lake Huron
Lake Michigan
Lake Ontario
Lake Erie

Only Lake Michigan is situated entirely in the U.S.A. The other four cross the border between the U.S.A. and Canada.

River facts

Mississippi, U.S.A. – 3,741 miles
Volga, Europe – 2,300 miles
Murray, Oceania – 1,609 miles
Chang Jiang, Asia – 3,900 miles
Amazon, South America – 4,000 miles
Nile, Africa – 4,160 miles

This diagram shows the comparative lengths of the longest rivers in each continent.

World's longest rivers

Name	Continent	Length (mile)
Nile	Africa	4,160*
Amazon	South America	4,000*
Chang Jiang (Yangtze)	Asia	3,900
Mississippi	U.S.A.	3,741
Huang He (Yellow)	Asia	3,000
Congo (Zaire)	Africa	2,900

*The Nile and the Amazon can both be measured from various points, making their official lengths vary.

Where would you find a delta?

Where rivers meet the sea, they can form v-shaped deltas. The river flows slowly and piles up silt into sandbanks, through which the river flows into the sea. The biggest delta is in Bangladesh, where the Ganges and Brahmaputra rivers create a delta almost as big as England. The Mississippi River delta in the United States extends over 180 miles into the Gulf of Mexico.

➲ *The river Nile empties its waters into the Mediterranean sea through a great delta.*

Can a river flow backward?

An incoming push of seawater, as the tide flows in, can make a wave that rushes upstream—so making a river flow backward. This wave is called a tidal bore, and there is a famous one on the Severn River in England, which moves upstream at about 12 miles an hour. A bore more than 20 ft. high rushes up the Qiantang River in China. Normally, rivers flow from their source (often on a mountain) downhill to the sea, drawn by the pull of gravity.

⬇ *The highest falls are the Angel Falls in Venezuela, South America, with one drop of 2,648 ft., and a total drop of 3,212 ft., more than twice as high as the Empire State Building! This diagram compares the largest waterfalls in each continent.*

⬆ *Victoria Falls are on the Zambezi River, on the border between Zambia and Zimbabwe in central Africa. Waterfalls with such immense flows of water produce huge clouds of spray and a thunderous noise.*

What makes a waterfall?

Waterfalls occur when a river flows over a band of hard rock, and then over softer rock which is more quickly worn away by the water. The hard rock forms a step over which the river pours, creating a waterfall. Famous waterfalls include Niagara in North America, Angel Falls in South America, and Victoria Falls in Africa.

Angel Falls
South America
3,212 ft.

Giessbach Falls
Europe
1,982 ft.

Sutherland Falls
Oceania
1,904 ft.

Ribbon Falls
North America
1,611 ft.

Jog Falls
Asia
830 ft.

Victoria Falls
Africa
354 ft.

Largest **lakes**

Name	Location	Area (sq miles)
Caspian Sea	Asia	143,200
Lake Superior	North America	31,700
Victoria Nyanza	Africa	26,600
Huron	North America	23,000
Michigan	North America	22,300
Tanganyika	Africa	12,600
Baikal	Russia	12,200

Longest **lakes**

Name	Length (miles)
Caspian Sea	746
Tanganyika	420
Baikal	395
Nyasa/Malawi	350
Superior	350

➲ *Glacier lakes are created by glacial erosion, which deepens the valley floor. Water fills these valleys and moraine debris falls from the glacier to form a dam to keep the water in.*

Viewed from space, the Earth looks like a planet of blue ocean—more than 70 percent is water. About 97 percent of all the Earth's water is in the oceans, which cover more than 140 million sq miles of the planet. The oceans can be more than 6 miles deep in places.

What causes tides?

Tides rise and fall twice every 24 hours, as the gravity of the Sun and Moon pull on the waters of the Earth—drawing the ocean toward them. The oceans move in different ways. Currents are streams of warm and cool water, some pushed by winds, others by the tides. The land is pulled too, but water moves more easily, causing a vast wave, which moves around the globe and forms the tides.

Spring tides (high)

As the tide rises, the sea flows upward and inland

Neap tides (low)

As the tide drops, the sea ebbs, retreating from the shore

How many oceans are there?

There are five oceans, which all connect to make one vast body of water. The three biggest oceans are the Pacific, Atlantic, and Indian Oceans. They meet around Antarctica, in the Southern or Antarctic Ocean. The Pacific and the Atlantic also meet in the smaller Arctic Ocean. Seas, such as the Baltic Sea, are smaller areas of saltwater, but most seas are joined to an ocean, such as the Mediterranean Sea, which is linked to the Red Sea by the Suez Canal, and to the Atlantic Ocean at the Strait of Gibraltar.

◔ *There are two high and two low tides every 24 hours. The very high spring tides occur when the gravity from the Sun and the Moon combines to tug at the oceans.*

Which is the biggest ocean?

The Pacific Ocean is the biggest of the Earth's oceans. About 45 percent of all the seawater on the Earth is in the Pacific Ocean. It has a surface area of 66 million sq. miles, which is equivalent to one-third of the Earth. The Pacific Ocean is larger than the second and third biggest oceans (Atlantic Ocean at 34 million sq. miles, and Indian Ocean at 26.6 million sq. miles) combined.

◔ *Some maps name the North and South Pacific, and the North and South Atlantic separately, but the Pacific and the Atlantic are each one ocean.*

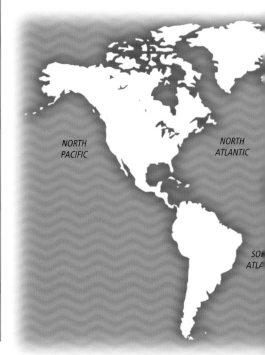

NORTH PACIFIC

NORTH ATLANTIC

SO ATLA

Seas and **oceans**

Seas, tides, and waves
Seas are named parts of oceans, often partly enclosed by land. The biggest sea is the South China Sea, which covers 1,300,000 sq miles.

Waves are blown along by winds. The highest wave ever seen was in 1933 when, during a Pacific hurricane, a U.S. Navy ship survived a wave estimated to be 112 ft. high.

The highest tides (more than 45 ft.) rise and fall in the Bay of Fundy, on the Atlantic coast of North America.

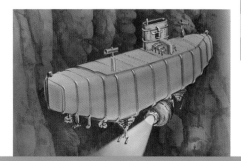

Ocean **depths**

Ocean	Deepest point
Pacific	35,840 ft.
Atlantic	28,232 ft.
Caribbean	24,720 ft.
Indian	23,812 ft.
South China	16,457 ft.

◔ *The Pacific has the five deepest trenches in the oceans. In 1961, the U.S. Navy bathyscaphe* Trieste *dove to the bottom of the Marianas Trench, the deepest point.*

What is it like on the ocean floor?

The ocean floor is a varied seascape of mountain ranges, deep trenches, hot springs, and mud oozing for hundreds of meters. At the coast, the land slopes gradually to a depth of about 600 ft.—this is the continental shelf. At the edge of the shelf, the ocean floor drops away in the continental slope, leading to the deepest part of the ocean floor, the cold and sunless abyss. On the ocean floor, hot springs reach temperatures above boiling point (212°F) in places.

◐ *The ocean floor has features, just like dry land, with ridges, canyons and mountains.*

Continental shelf

Continental slope

Ocean trench

Why is the Dead Sea so salty?

The saltiest sea is the Dead Sea, which is enclosed by hot desert where the fierce heat evaporates so much water that what is left becomes very salty. Seawater tastes salty because it contains minerals washed into the oceans by rivers. The most common mineral in seawater is common salt (sodium chloride).

Could you drink a melted iceberg?

Yes, because although icebergs drift on the ocean, they are made of fresh water. Icebergs break off the ends of glaciers— slow-moving rivers of ice—that move down mountain slopes in the polar regions. Icebergs also break off from the edges of ice sheets.

◐ *Only about 11 percent of a tall iceberg shows above water; the rest is submerged.*

◐ *The water of the Dead Sea is so salty that it is more buoyant than ordinary seawater.*

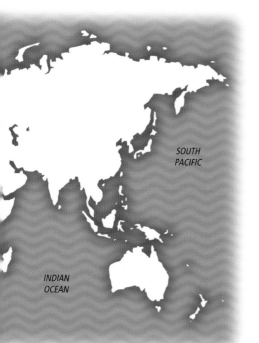

SOUTH PACIFIC

INDIAN OCEAN

Ocean **trenches**

The longest and deepest ocean trenches are:

Name	Length (miles)	Depth (ft.)
Peru-Atacama	2,199	29,738
Aleutian	1,988	26,574
Tonga-Kermadec	1,600	35,702
Marianas Trench	1,398	35,797
Philippine	823	34,438

◐ *Many islands in the Pacific Ocean, including those of Japan and Hawaii, were formed by volcanoes, which pushed up islands in "chains."*

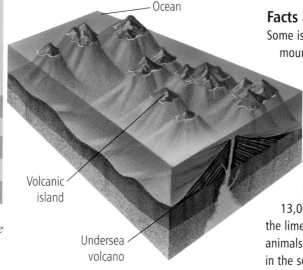

Ocean

Volcanic island

Undersea volcano

Facts about islands

Some islands are the tops of undersea mountains or are pushed up by volcanoes. Other islands (for example, the British Isles) were once part of continents but became cut off by water. A chain or group of islands is called an archipelago. The world's biggest archipelago is Indonesia with more than 13,000 islands. Reefs are made of the limestone bodies of tiny coral animals, which form rings and walls in the sea.

Dry desert, where the land gets less than 10 in. of rainfall in a year, covers almost one eighth of the Earth's surface. The driest place on Earth is the Atacama Desert in Chile, South America, where many years pass without a single drop of rain. Not all deserts are hot though. The hottest ones are near the equator, but even there it can be chilly at night.

➡ *People and animals have adapted to desert life. Arabian camels are still used as baggage carriers by desert nomads, who traditionally traveled from oasis to oasis, living in tents that provided shade by day and warmth by night (when it gets cold in the desert).*

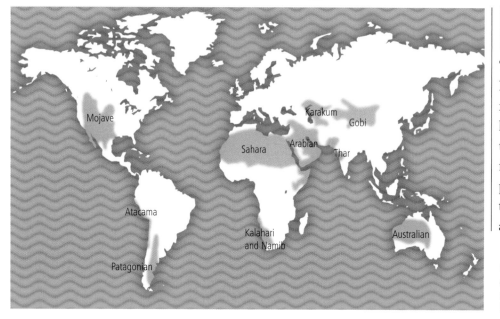

Which is the biggest desert?

The biggest desert is the Sahara in North Africa, which is more than 3,500 miles across, bigger than Australia. It has the world's largest sand dunes, some of them more than 1,300 ft. high. Cave paintings found near the region, drawn by ancient people, depict grassland animals. This shows that thousands of years ago the Sahara was actually wetter, with lakes and plains.

⬅ *Dry deserts occur in warm areas where cool air sinks, warms up, and then absorbs moisture from the land. The map shows the world's biggest deserts.*

Deserts and **desertification**

Desert environment
People on the edges of deserts are living in a very delicately balanced environment. Water is precious, and so are the trees and other plants that bind the thin soil and help hold back the desert sands. Careless farming, or too many people trying to scratch a living from the desert fringes, can turn half-desert into full-desert. This is called desertification. Farmers who keep too many sheep and goats, or cut too many trees for fuelwood, find that the desert is encroaching onto once fairly fertile pastures and fields. Climate change can also cause deserts to expand. Deserts seem to ebb and flow, like the oceans. Satellite photos of the Sahara, for example, show that it spreads southward for some years, then shrinks. About 450 million years ago, the Sahara was covered in ice. In the Sahara are huge seas of sand, called ergs. The biggest erg, between Algeria and Tunisia, covers an area as big as Spain.

1 Sahara Desert

Are all deserts sandy?

Only about 20 percent of the Earth's deserts are sandy. The rest are rocky, stony, covered by scrub and bush, or ice-covered. In the Arabian Desert is the world's biggest area of sand dunes—the Rub' al-Khali, which means "empty quarter" in Arabic.

⬆ *Sand dunes can travel across the desert, like waves across an ocean.*

Can sand dunes move?

Loose sand is blown by the wind and piles up in wave-shaped formations called dunes. Sand is made up of tiny mineral grains, less than 0.08 in. across. Like waves of water, sand is blown up, rolls over the crest of the wave and down the steeper far side. Dunes rolls across the desert in this way.

⬆ *Underground water allows people to cultivate palms and vegetables in a desert oasis. Some oases support small towns.*

What is an oasis?

An oasis is a green "island" in the desert, a haven for thirsty travelers. Plants can grow there by tapping water from a well or underground spring. Even beneath the Sahara Desert a lot of water is trapped deep in the rock strata (layers).

How do desert animals and plants survive?

Desert animals are able to go for days without water, getting most of the moisture they need from their food. These animals include mammals such as antelope, camels, foxes, and rodents, as well as birds and insects. Other animals, such as desert frogs, go into a state of suspended animation in burrows until the next rain.

⬇ *Many desert creatures, such as insects, scorpions, lizards, and snakes feed at night, when it is cooler.*

⬇ *The largest desert in the world is the Sahara, followed by the Australian desert. Asia has more deserts than any other continent. They include the Gobi, Karakum, Taklimakan, and Thar deserts.*

2 Australian Desert
3 Arabian Desert
4 Gobi Desert
5 Kalahari Desert

Biggest **deserts**

Name	Location	Area (sq miles)
Sahara	North Africa	3.47 million
Australian	Australia	1.46 million
Arabian	Southwest Asia	1 million
Gobi	Central Asia	500,000
Kalahari	Southern Africa	190,000

➡ *This tree in the Namib Desert of southwestern Africa manages to survive, even in such a hostile, dry environment.*

Many trees covering an area of land create a forest. A wood is a smaller area of trees. Some forests in cooler regions of the Earth have only two or three species of trees growing in them, but other forests have a tremendous variety of trees and other plants. The first forests grew in prehistoric swamps over 350 million years ago, but these were forests of tall ferns and moss-like plants, not trees.

Are there different types of forest?

There are different types of forest around the world, depending on the climate in the world's vegetation zones. In the warm tropics there are rain forests and seasonal forests (where trees lose their leaves during the dry season) and savanna (warm grassland) forests. Rain forests also grow in cooler zones, where there is a lot of rainfall. In cooler zones, there are forests of mixed deciduous trees (which shed their leaves before winter), and of evergreen conifers, such as fir and pine. Boreal forest or taiga is found in cold subpolar lands.

Which is the world's biggest rain forest?

The Amazon rain forest of South America, which stretches from the foothills of the Andes Mountains in the west to the Atlantic Ocean in the east.

↑ *About 40 percent of the the world's plants are found in rain forests.*

There are other rain forests in west Africa, Southeast Asia, and northeastern Australia. Rain forests are abundant with wildlife. There are more species of animals and plants in the Amazon than anywhere else on Earth.

What starts a forest fire?

Forest fires can begin naturally, when the vegetation is very dry after months without any rain. Often humans are responsible for starting the fires by carelessly lighting campfires or sometimes even by deliberate arson. Many forest trees and other plants regenerate quickly after a fire, but wildlife can be seriously affected.

⊕ *Forest fires can spread rapidly. Firefighters often cut trees to create a strip of open ground, called a firebreak, to stop the flames spreading.*

Forest **facts**

Main canopy

Emergent tree

Understory

Shrubs

⊕ *This cutaway of a rain forest shows the canopy. Trees compete to reach the sunlight above.*

Amazing **forest facts**

- Many forest trees grow to great ages if they are left undisturbed by people. Oaks of 500 years old are not uncommon in ancient forests.

- Just 2.5 acres of rainforest can contain 180 tree species.

- A rainforest has different levels, like floors in a building. The thickest part is the main canopy, about 100 ft. high, where most animals live. Taller trees emerge from the canopy.

- Tropical forests grow densely because the trees get sunlight and rainfall almost every day. It is a year-long growing season, so plants grow very fast.

↥ *There are 450 different species of oak trees most of which belong to the* Quercus *group, which grow mostly in the northern parts of the world. Oak trees can live for more than 1,000 years and some of the oldest trees in Europe are oaks. Wood from the oak tree is very strong and has been used in the building of houses and ships over the past centuries.*

Why do some forest trees shed their leaves?

Trees in a deciduous forest shed their leaves to save water, because their roots cannot soak up water very well from cold soil.
Deciduous forests grow in countries with warm summers and cool winters. The trees in these forests include oak, beech, maple, ash, and chestnut, and as summer gives way to autumn, the leaves of the trees change color and begin to fall.

What is meant by a sustainable forest?

Forests provide people with many products, including timber, foods, cosmetics, and drugs. Coniferous forests of pine, spruce, and fir are felled for timber or newspaper pulp. In a well-managed sustainable forest, new trees are planted to replace those that are felled. Sadly, many tropical rain forests are being destroyed not just for timber, but also to clear the land for farming and ranching. Felled trees are not replaced, leaving a stump-littered wasteland.

Natural forest

Trees are burned

Farm crops

Rain washes away topsoil

Soil becomes useless

↥ *Widespread deforestation can have devastating effects on the landscape. Tropical soil rapidly loses fertility. Rain washes off the topsoil, and once-lush forest becomes scrubland or desert.*

- Tropical rain forests grow luxuriantly because the rainfall is heavy and regular—often more than 80 inches of rain in a year.

- South American and Indonesian rain forests are most in danger from exploitation. Every year the Amazon rainforest loses about 31,000 sq. miles of trees.

- Some of the world's tallest trees grow in Australia's eucalyptus forests. Karri, mountain ash, and blue gum trees grow to more than 130 ft. high.

- Some trees, such as the Douglas fir, are tough enough to survive even on cold and windy mountains. Conifers withstand the cold better than deciduous trees, but the higher a mountain is, the more difficult it is for trees to survive. Above the tree line is bare mountainside.

↧ *Managed forestry means replacing trees that are cut by loggers (left) with new saplings (young trees) that will grow and insure the forest survives.*

The atmosphere is the layer of gases that surround the Earth. It is held in place by the Earth's gravity, which keeps most of the gases in the atmosphere close to the ground. Most of the atmospheric gases are packed into the lowest layer of the atmosphere, starting at ground level, which is called the troposphere.

Exosphere 310–500 miles

Thermosphere 50–310 miles

Mesosphere 30–50 miles

Stratosphere 6–30 miles

Ozone layer

Troposphere 0–6 miles

We live in the lowest layer of the atmosphere. Planes cruise in the layer called the stratosphere. Phenomena such as the auroras—Borealis of the northern hemisphere and Australis in the southern hemisphere—occur in the thermosphere. Above that lies space.

How many layers are there in the atmosphere?

There are five main layers in the atmosphere. The lowest is the troposphere, up to 6 miles high. Next is the stratosphere, about 30 miles high, and above that is the mesosphere, to about 50 miles. The upper layer is called the thermosphere. The higher you go, the thinner the atmosphere, and above 500 miles there is no atmosphere left and the exosphere (a very thin fifth layer) gives way to the airless, near emptiness of space.

How does the atmosphere protect us?

In the upper levels of the atmosphere is a layer of ozone (a form of oxygen), which forms a protective layer blocking out harmful ultraviolet rays from the Sun. On the fringes of the atmosphere are two doughnut-shaped radiation belts, known as the Van Allen belts, which shield us from cosmic rays coming from space.

The atmosphere is a protective belt, burning up meteorites and shielding life on Earth from harmful cosmic radiation.

Atmospheric **information**

In the atmosphere
The atmosphere contains oxygen, nitrogen, and tiny amounts of other gases, argon, carbon dioxide, carbon monoxide, hydrogen, ozone, methane, helium, neon, krypton, and xenon. The most important gas in the atmosphere is oxygen, because people and animals need to breathe it. When we breathe, we take in oxygen and breathe out carbon dioxide. Green plants, such as trees, take in carbon dioxide and give off oxygen during their food-making process (photosynthesis).

At sea level, one cubic foot of air weighs less than 1.2 ounces. Imagine the weight of air pressing down on top of your head—it comes to almost a ton. Luckily we do not feel crushed, because the pressure is balanced by the air pressing in all around us. At sea-level pressure is about 14.7 pounds per sq. in. The pressure drops with altitude, so at 5,000 ft. the pressure is only 1.5 pounds per sq. in. The higher you go, the colder it gets, too. At a height of 30,000 ft. the air temperature is –58°F.

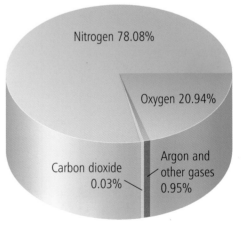

Nitrogen 78.08%

Oxygen 20.94%

Carbon dioxide 0.03%

Argon and other gases 0.95%

Nitrogen and oxygen together make up about 99 percent of the Earth's atmosphere.

Why does the sky look blue?

Light from the Sun passes through the atmosphere and is scattered by tiny particles of dust and moisture in the air. This has the effect of breaking up the white sunlight into its rainbow colors, just like a spectrum does. The blue rays scatter most, and reach our eyes from all angles. The result is that we see blue more than the other colors in sunlight, so the sky looks blue.

⬆ *The northern lights, or aurora borealis, makes the sky glow and streak in a beautiful display of colors ranging from purple to golden-green.*

⬆ *Clouds change shape all the time. Some are big billowing masses, others are feathery traces. The wind changes the shape of clouds, which often obscure the blue sky above.*

What causes the northern and southern lights?

The northern and southern lights are caused by the solar wind (radiation from the Sun) hitting the atmosphere. Most of the incoming energized particles are absorbed by the Van Allen belts, but at the poles, where the belts are thinnest, the particles impact on the atmosphere, producing a spectacular light show in the sky. The aurora borealis (northern lights) makes the night sky glow green, gold, red, or purple. The aurora seen in the southern hemisphere is the aurora australis (southern lights).

Where is the air coldest?

Anyone who climbs a mountain soon realizes that it gets colder the higher you climb. The temperature falls by about 4°F for every 1000 ft. altitude. However, in the atmosphere high above ground level, it is a different story. The outer layer of the atmosphere can be warmer than the layers closer to the surface. Within the layer called the troposphere it gets coldest when the air rises highest, which is over the equator. It is actually warmer over the north pole, where the air does not rise as high.

Air-pressure systems

Two main air-pressure systems control our weather. High pressure (anticyclones) form when cold air sinks. High pressure usually means fine, dry weather—warm in summer, cold in winter. Low pressure (cyclones or depressions) occur when warm air rises, bringing rain clouds and unsettled weather. Winds blow from high pressure zones into low pressure zones. The strength of the wind depends upon how great the difference in pressure is. If there is a big difference, then the wind is strong.

⬅ *In the northern hemisphere, winds spiral clockwise from high pressure zones and counterclockwise into low pressure zones.*

➡ *In the southern hemisphere, winds spiral in the opposite direction to those in the northern hemisphere. They spiral clockwise into low pressure zones.*

Regular weather charts have been kept for only about the last 250 years, and accurate temperature readings date from the 1800s. But people have always been interested in the weather. Historic Chinese weather records show that 903 BC was a very bad winter in China, and the Romans recorded that the weather was poor when they landed in Britain in 55 BC.

How deep can snow be?

The most snow to fall in 24 hours was 6.3 ft.—enough to cover a tall man. This snowfall buried Silver Lake in Colorado, between April 14 and 15 1921. The deepest snow covering ever measured was 37.5 ft. in California in 1911—enough snow to cover a small house to the roof!

Snowfall tends to be heaviest in mountainous regions, such as the Rocky Mountains and Sierra Nevada ranges of North America. People who live in these regions have learned to cope with blizzards (heavy snowfall) that half-bury cars and houses.

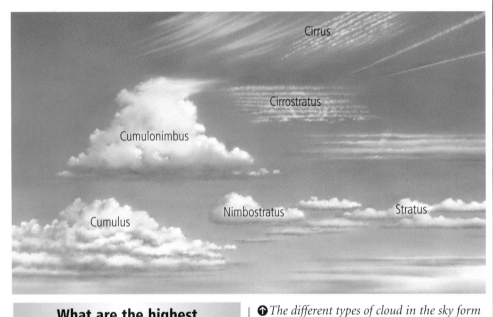

What are the highest clouds in the sky?

The highest clouds are the rare "mother of pearl" nacreous clouds, which can be found at 80,000 ft. Cumulonimbus clouds can tower as high in the sky as 60,000 ft. The more common cirrus clouds form at around 26,000 ft. The lowest clouds are stratus clouds, from 3,600 ft. to ground level.

The different types of cloud in the sky form at different heights. Clouds are made of tiny droplets of water or ice.

Snowflakes are made up of snow crystals, that can be seen under a microscope. Each has six sides but every snowflake is different.

Weather **facts**

 Occluded front—where a cold front (blue triangle) meets a warm front (red semicircle)

Strength of the wind—the circle shows how much cloud cover there is

 Very strong winds—evident from the three lines on the tail of the symbol

 Temperature at sea

Air pressure

Cloud cover shown by circle

Center of low air pressure

Meteorologists (scientists who study the weather) use an international set of symbols to represent different aspects of the weather.

Stormy **weather**

- In the U.S.A., tornadoes are called "twisters" and roar across the Midwest at speeds of 30 mi./hr.

- Britain's worst storm of modern times was the hurricane of 1987, which blew down about 15 million trees in southern England.

- Windspeeds are measured on a scale invented in 1806 by an English admiral, Sir Francis Beaufort. The scale goes from 0 (calm) to 12 (hurricane). Any winds stronger than force 8 can cause damage.

Where is the thickest ice?

The ice covering Greenland is about 1 mile thick, but the ice in Antarctica is three times thicker, up to 3 miles thick! Antarctic icebergs are flatter than Arctic ones. The biggest iceberg was spotted in the Antarctic in 1956. It was 208 miles long and 60 miles across. A country the size of Belgium would have fitted on top of it.

⬆ *Lightning flashes between clouds, or from cloud to ground, heating the air around it to more than 60,000°F. This is five times hotter than the surface of the Sun.*

What is a hurricane?

The most destructive storms are hurricanes, known as cyclones in the Indian Ocean and typhoons in the Pacific Ocean. In a hurricane, winds spiral at speeds of more than 350 mph, yet at the center is a calm area, known as the "eye" of the hurricane. A hurricane does most damage when it hits the land.

What causes lightning?

A lightning flash is a giant electric spark, caused by electrical charges that build up inside clouds and on the ground. Lightning is incredibly hot and so can seriously injure and even kill people. An American park ranger named Roy Sullivan was struck by lightning seven times (and survived) between 1942 and 1977.

⬅ *Hurricanes can be photographed from high above, in space. The 'eye' can be seen quite clearly on the photographs. Satellites in space track hurricanes over the ocean.*

➡ *Each color in the white sunlight is bent to a different extent. The light is split into the spectrum colors—red, orange, yellow, green, blue, indigo, and violet—to form the rainbow.*

When might you see a rainbow?

You might see a rainbow—an arc of up to seven colors in the sky—during a rainfall. A rainbow is caused by light being refracted (bent) by the raindrops. To see a rainbow, the Sun must be behind you. If the Sun is high in the sky, no rainbow will appear.

Extreme **weather**

Most thunder—Tororo in Uganda, Africa, has an average of 250 thundery days every year.

Worst hailstorm—A hailstorm in 1888 killed 246 people in India.

Biggest hailstone—Bigger and heavier than a tennis ball, at 17.5 in. around and weighing 2.2 lb., it fell in Kansas, in 1970.

Highest waterspout—A waterspout reported to be 1,500 m high was seen off the coast of New South Wales, Australia, in 1888.

Worst cyclone—In 1991, 138,000 people were killed when a cyclone and tidal wave hit Bangladesh.

Hottest place—In Death Valley in California, USA, the thermometer stayed above 120°F for 43 days in 1917.

⬆ *Hailstones measuring up to 2 in. can fall during a hailstorm. A hailstone with a diameter of 7 in. was recorded in the U.S.A.*

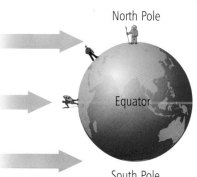

⬆ *The poles are colder than the equator because they receive less direct sunlight. At the poles, the Sun's rays have more atmosphere to penetrate, and so lose some of their warmth.*

The Earth has many natural resources, which sustain life. The planet has air, water, forests, minerals, and a range of environments that living things can use. Some of these resources, such as the Sun's energy, are limitless. Others are renewable, such as plants, which means that they can be regrown. But some, like coal and oil, are nonrenewable resources. Once used, they are gone for ever.

⬅ *Oil is an extremely valuable natural resource that we use in our homes and for transportation. It is found deep in the ground or below the seabed, pumped out by oil rigs that float in the sea, anchored to the seabed.*

⬆ *Diamond is the hardest known substance.*

Why is a diamond like a lump of coal?

Diamond and coal are both forms of carbon. About 95 percent of all compounds (substances made of two or more elements) contain carbon, which is the key element in substance building because it has atoms that form chains, rings, and other structures, making them stronger and more durable.

⬇ *A hydroelectric power station uses water to drive its turbines.*

How is water used to make electricity?

Water is stored in a vast dam, and runs through pipes at great speed to drive turbines that generate electricity. An electricity generator works by turning motion (one form of energy) into electricity (another form of energy). This motion may come from a turbine that is driven by steam or water.

What are raw materials?

They are the Earth's resources that make our lives more comfortable. We cut trees for timber to make homes and furniture. We mine minerals, such as copper, to make the electrical wire in our buildings. We mine coal to burn as a fuel. All of these materials are nonrenewable, which means we cannot make more of them.

Clean energy **and recycling**

Energy

Wind power is one method of generating electricity without using up precious resources. The wind is a renewable resource because in many parts of the world it seldom stops blowing. Wind turbines with spinning blades are grouped in wind farms, which supply electricity to the grid system.

Other forms of renewable clean energy are wave and tidal power, which harness the motion of waves; and solar power, which turns the energy from the Sun into electricity.

In future, people will rely on a mix of these energy sources, but it is very important today to save energy and to recycle the things we use. Much of our "garbage" can be recycled—used again—in a variety of ways. Materials, such as glass, paper, and metals can be treated and processed in a factory to be reused in the same form.

⬅ *Wind farms are usually in remote areas or offshore, where the noise of the turbine blades causes less disturbance to people.*

What is coal?

Coal was formed over 250 million years ago, from dead and decaying plants in prehistoric swamp forests. Over that vast length of time, the plant matter got squashed so tightly that it changed and became a soft black-brown rock. Coal is found in seams or layers, with other rocks on top and beneath. It is extracted either by digging deep mine shafts and tunnels, or by "stripping" coal seams near the surface, known as opencast mining.

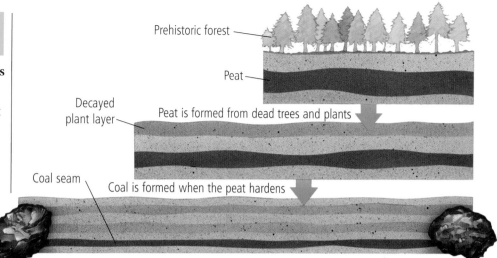

Prehistoric forest

Peat

Decayed plant layer

Peat is formed from dead trees and plants

Coal seam

Coal is formed when the peat hardens

Where is the world's richest goldfield?

The mines of the Witwatersrand in South Africa yield about 50 percent of the world's annual gold production, making them the richest gold mines in the world. Several times in history gold miners have rushed to find gold in various parts of the world. The most famous gold rush was in California in 1849. There was another in South Africa in the 1880s.

What is the "greenhouse" effect?

Gases, such as carbon dioxide, act like the glass in a greenhouse, letting through the Sun's rays, but trapping some heat that would otherwise filter back into space. Since the 1800s, human activity (especially factories, vehicles, and power stations) has caused an increase in the amount of carbon dioxide and other gases in the atmosphere. The trapped warmth that is created raises the Earth's temperature and many scientists believe this is bringing about climate change.

⬅ *Few gold-rush prospectors ever made the fortune they dreamed about by finding nuggets of gold.*

⬆ *This diagram shows how coal seams are formed over millions of years. Much of the coal lies deep beneath rock layers, called strata. The pressure of the topmost layer squeezes the layers below, turning sand and mud into hard rock, and plant remains from peat into coal.*

⬇ *Gases that have risen into the atmosphere trap the heat from the Sun, causing what scientists call the "greenhouse" effect.*

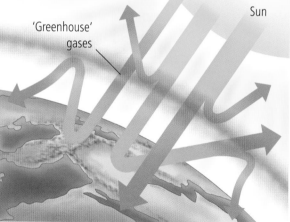

'Greenhouse' gases

Sun

Recycling **tips**

Cardboard boxes—can be recycled to make packaging.
Newspapers—can be recycled to make more paper.
Glass—can be melted down and reformed.
Metal cans—and other metal scrap can be melted down.
Some plastics—can be shredded and reused.

Many towns and cities now operate recycling schemes for household rubbish.

⬆ *Recycling saves precious resources and helps keep our environment "healthy," avoiding an accumulation of household garbage, which pollutes the environment.*

Save **energy**

- Buy products with the least packaging.
- Turn off electric lights and appliances after use.
- Keep car use to a sensible minimum to save fuel and air pollution.
- Recycle household waste, such as metal, paper, plastic, and glass.
- Do not leave faucets running.
- Turn heating down or off when the weather is mild.

Why not test your knowledge of planet Earth! Try answering these questions to find out how much you know about volcanoes, earthquakes, maps, rivers, lakes and oceans, forests, deserts, continents, and much more. Questions are grouped into the subject areas covered within the pages of this book. See how much you remember, and discover how much more you can learn by looking at other sources to help you answer these questions.

Inside the Earth

15 The inner core of the Earth acts like a huge magnet: true or false?
16 The world's deepest mine can be found in which country?
17 Is the mantle below Earth's crust mainly solid or molten rock?

Earth Facts

1 Is there more land or water making up the Earth's surface?
2 How long does it take for the Earth to revolve around the Sun?
3 Is the circumference of the Earth 249 miles, 2,490 miles or 24,902 miles?

Rocks and Fossils

4 Is chalk a type of limestone or granite rock?
5 Geothermal energy comes from what source?
6 What type of rock was marble before it was changed under great pressure?

7 What type of scientists study fossils, such as this ammonite?

8 India and Asia have pushed together over the last 40 million years to create which mountain range?

Continents

9 Which continent has the highest population?
10 What are the remote areas of Australia's interior known as?
11 On which continent would the Rhine River be found?

Maps and Globes

12 What is made during the process of cartography?
13 Do longitude lines run horizontally or vertically around the Earth?
14 At what degree latitude is the equator: 0°, 45° or 90°?

Volcanoes and Earthquakes

18 The three most active volcanoes in South America are found in which country?
19 What rock formed by volcanoes can be so light it will float on water?
20 What is the name given to a small warning shock delivered by an earthquake?

Landscapes

21 Is a plain a flat or hilly area?
22 Is four fifths of the world's farmland used to feed people or animals?
23 The Grand Canyon is found in which country?

Rivers and Lakes

24 The Thames River flows through which British city?

25 In which continent would you find the Amazon River?

26 On which river are the Victoria Falls?

Oceans

27 What ocean separates Europe from North and South America?

28 What is the world's largest ocean?

29 Which ocean is larger: the Indian or the Atlantic?

Deserts

30 What is the largest desert in Africa?

31 The Gobi Desert can be found on which continent?

32 Where is the Great Victoria Desert?

Forests

33 Are pine trees hard or softwoods?

34 In which continent is the world's largest rain forest?

35 Rubber can be made from the sap of certain trees: true or false?

36 Italian Evangelista Torricelli studied under Galileo and was the first to make what type of meteorological instrument?

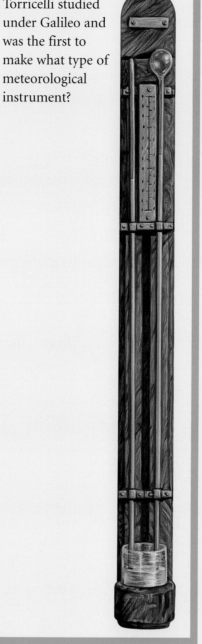

Atmosphere

37 What is the second most common element in the air?

38 What is mixed with fog to make smog?

39 In which layer of the atmosphere do airplanes cruise?

Weather and Climate

40 What simple device can show the direction of the wind?

41 If you were in the northern hemisphere in October, what season would you be in?

42 Meteorology is the study of what feature of Earth?

Earth Resources

43 What is amber: a mineral, a flower, or a metal?

44 Is oil a fossil fuel?

45 Tenant Creek and Weipa in Australia are famous mining areas for which precious metal?

Answers

1 More water	13 Vertically	25 South America	37 Oxygen
2 One year	14 0°	26 Zambezi	38 Pollution
3 24,902 miles	15 True	27 Atlantic	39 Stratosphere
4 Limestone	16 South Africa	28 Pacific	40 Wind vane
5 Heat from the Earth's rocks	17 Mainly solid	29 Atlantic	41 Autumn
6 Limestone	18 Chile	30 Sahara	42 Its atmosphere and weather
7 Geologists	19 Pumice	31 Asia	43 Mineral
8 The Himalayas	20 Foreshock	32 Australia	44 Yes
9 Asia	21 Flat	33 Softwoods	45 Gold
10 The outback	22 Animals	34 South America	
11 Europe	23 U.S.A.	35 True	
12 Maps	24 London	36 Barometer	

The publishers would like to thank the following artists who have contributed to this book:
C. Buzer/Galante Studio, Kuo Kang Chen, Mark Davis, Nick Farmer, Chris Forsey, Terry Gabbey, Luigi Galante,
Alan Hancocks, Richard Hook, Rob Jakeway, Maltings, Janos Marffy, Martin Sanders, Mike Saunders,
Guy Smith, Brak Syd, Rudi Vizi, Steve Weston, Paul Williams

The publishers wish to thank the following sources for the photographs used in this book:
Roger Ressmeyer/CORBIS p18 (c/r)

All other photographs are from:
Corbis, Corel, PhotoDisc